The Story of Science

The Ever-Changing Atom

by Roy A. Gallant

BENCHMARK BOOKS

MARSHALL CAVENDISH
NEW YORK

For Gary

Series Editor: Roy A. Gallant

Series Consultants:

LIFE SCIENCES
Dr. Edward J. Kormondy
Chancellor and Professor of Biology (retired)
University of Hawaii—Hilo/West Oahu

PHYSICAL SCIENCES
Dr. Jerry LaSala, Chairman
Department of Physics
University of Southern Maine

Benchmark Books
Marshall Cavendish Corporation
99 White Plains Road
Tarrytown, NY 10591-9001

Library of Congress Cataloging-in-Publication Data
Gallant, Roy A.
        The ever-changing atom/ by Roy A. Gallant.
        p. cm. — (The story of science series)
Includes bibliographical references and index.
Summary: Introduces atoms, the tiny particles which make up everything in the world, discussing their different parts, how they were discovered, and how they can be used as a source of energy.
ISBN 0-7614-0961-0
        1. Atoms—Juvenile literature. [1. Atoms. 2. Atomic theory 3. Nuclear physics.] I. Title. II. Series: Gallant, Roy A. The story of science series.
QC173.16.G35   1999   539.7—dc21   35420   CIP   AC

Diagrams on pp. 10, 12, 14, 19, 21, 22–23, 24, 27, 35, 36, 39, 46, 48, 56, 59, 73,
        by Jeannine L. Dickey

Printed in Hong Kong
6 5 4 3

Cover: Modern scientists learn about what atoms are made of and how their many bits and pieces work by making atoms smash into each other. The particle tracks shown on the cover resulted from the collision of two protons with two antiprotons. As the whorls of a finger print identify their owner, the tracks identify certain particles of an atom and how they behave.

Photo reseach by Linda Sykes Picture Research, Hilton Head, SC
Cover photo: © Laurence Radation/Photo Researchers
11, Derby Museum and Art Gallery, Derby, England/Bridgeman Art Library; 16 (bottom) Gary Retherford/Photo Researchers; 16 (top), Stock Montage; 18, Photo Researchers; 21, 27, Brown Brothers; 28, Corbis Bettmann; 130, Daniel Westergren/National Geographic Society Image Collection; 33, JL Charmet/Photo Researchers; 49, Martin Dohrn/Photo Researchers; 51, Los Alamos Laboratory/Photo Researchers; 53, The Library of Congress/Corbis; 60, Joseph Sohm, ChromoSohm Inc./Corbis; 67 Emory Kristof/National Geographic Society Image Collection; 68 Raymond Meier; 70 Dan McCoy/Rainbow; 72 Corbis Bettmann

Contents

One

What Is the World Made Of?

Let's Start with "Matter"

For a moment, forget what you think you know about matter. Try to imagine yourself as a scientist who lived 2,500 years ago, except there weren't any "scientists" then. In fact, there weren't even any of the sciences we know today—chemistry, physics, biology, or geology. (The science of astronomy was just being born.)

The puzzle you are trying to solve is to explain what matter is. What is the world "stuff"? Maybe it's made up of particles too tiny to be seen. But if you group enough of these tiny particles together, the bunch may become big enough to see. If so, then what are those invisible particles? Is there only one kind of particle, or are there many kinds? These are good questions, at least for a start.

Here's another one: If matter is made of tiny invisible particles, what holds the particles together, as a rock, for instance? What's the sticking force? And how can some particles slip and slide over one another, as particles of water do? Since water flows, the sticking force of its particles seems to be much weaker. And how can particles that make up the air stay so far apart from one another? Maybe they don't have any sticking force at all. You can tell that air is made up of *something*, because you can feel it push against your face when the wind blows.

But where do you go from there? Matter is "something," and it has some sort of sticking force.

Good Morning, Mr. Thales

To find out how the great thinkers of long ago regarded matter and its binding forces, let's begin by moving back in time and visiting one of them. We'll turn the clock back to around 600 B.C., some 2,600 years ago, and knock on the door of the ancient Greek philosopher Thales. Thales helped set up the world's first scientific school in the Greek city of Miletus. He greets you:

THALES: Good morning, and what a lovely morning it is. How can I help you?

YOU: I've just traveled through time from the future and have come to ask you about your ideas of matter and what holds it together. We know a lot about that stuff today, in my time, I mean. But I want to find out how you and the other Greek guys—ah, excuse me, *philosophers*—started the ball rolling. I mean how the history of matter got started, you know?

THALES: Well, yes, of course. Your quest is a worthy one. As one of your philosophers—ahem, I mean *scientists*—as you call them, once said. "The key to the present is the past." So you have done well to travel back in time to visit our school.

YOU: Is it okay if I turn on my tape recorder?

THALES: I don't know what a tape recorder is, but if it's some kind of transcribing device and won't blow up my school, it's "okay," as you put it. Well, to start with, my thinking about matter was the most important, most profound, for it was I, after all, who established the foundations of the inquiry into the composition and structure of matter, or "started the matter ball rolling," as you might put it.

My idea came to me after much thinking and observing. It would have been much easier to say what matter is if there were only one kind. But there are so many different kinds of matter—copper, wood, liquids, air, mist—the problem was not easy. It was difficult because one basic kind of matter had to be able to change and form all the other very different kinds of matter.

YOU: Hey, Mr. Thales, that's cool, er, I mean profound. So what is that basic kind of matter? I guess that's what I really want to know. Then I won't bother you any more.

THALES: Water.

YOU: What? Only water? That's *all*? Whatever gave you that idea? Don't get me wrong, I like it. It's real cool.

THALES: Exactly. Water can be cool. It can be ice. It can also be warm or hot. It can be steam. I'm delighted to see that you grasp my point so quickly.

YOU: Er . . . sure.

THALES: Those changing qualities of water allow it to exist as vapors, such as your breath. As liquids, such as wine and honey. And as solids, such as iron and this building of stone. Water, in its different forms, can account for everything there is.

7

YOU: I see. I mean, I don't *really* see. In my science class, Mr. Thompson says that an idea in science has to be able to be tested. Can you test your idea? I mean, *have* you tested your idea?

THALES: Of course not. I mean no disrespect to your Mr. Thompson, but there is no need for a test. We think logically and form our ideas by observing things around us. The importance of my idea is not whether it can be tested. Its importance is that I have proposed that but a single substance makes up all matter, whether living or nonliving.

YOU: Yes, but—

THALES: I'd rather not go into more detail right now. I have a student waiting. Thank you for coming to visit, and good luck with your quest.

(You thank Mr. Thales, put your tape recorder back in its case, and return to the present, a bit bewildered.)

Two

Earth, Air, Fire, and Water

The Elements of Empedocles

Around 450 B.C., about 150 years after Thales lived, another Greek philosopher tried to explain the nature of matter, also without experimenting. His name was Empedocles. He said that all things were made of not one, but four basic substances. Those substances came to be called "elements," but not the elements we know today. They were earth, air, fire, and water. Trees, buildings, and clouds—all things—he said, had no substance of their own. They were only changing mixtures of earth, air, fire, and water. These four elements, said Empedocles, had four qualities—cold, wet, hot, and dry. Any two of these qualities when joined produced an element. For instance, wet and cold combined to produce water. Wet and hot made air. Hot and dry made fire. And dry and cold made earth.

Our bodies, according to Empedocles, were made of the four elements. Our bones were supposed to be made of one-half fire, one-quarter earth, and one-quarter water. The four-elements idea of Empedocles seemed sound enough to be widely accepted for some two thousand years. To this day certain expressions we use remind us of Empedocles. We speak of a friend's "fiery" temper, or on a stormy night we go out into the "elements." Or we may call an unsound idea "airy."

Empedocles's four elements were to flame the hopes and dreams of the old alchemists. These people practiced their chemical art in Europe for some five hundred years, well into the 1700s. They believed they could change copper into gold and lead into silver. All they had to do was find the right substance to mix with the copper. The substance would then change the "form" of copper into the "form" of gold.

Such far-fetched notions lingered on until someone came up

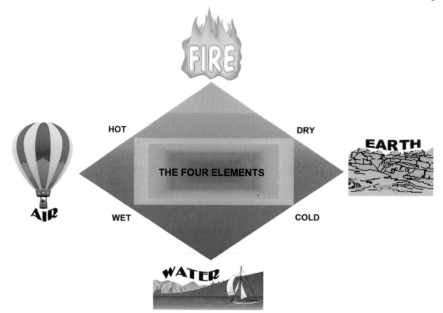

Before the true chemical elements were discovered, scientists of old imagined that all matter was mixtures of earth, air, fire, and water. The four qualities of hot, dry, cold, and wet could also be accounted for. Earth and water in combination produced the quality of cold. Fire and air combined to produce the quality of hot, and so on.

During the centuries before the true nature of the chemical elements was understood, experimenters called alchemists abounded. They believed that metals such as lead and mercury could be changed into gold. This painting, the Alchemist, *was made by Joseph Wright in the 1700s.*

with the idea that matter was made of tiny invisible particles—an atomic theory of matter. Particles of one kind could join with different kinds of particles and so form everything around us—salt, rust, glass, paper, and fillings for your teeth, for example. That someone was a Greek thinker named Democritus. He lived around 460–370 B.C.

The Atoms of Democritus

The main part of Democritus's atomic theory came from his teacher, Leucippus. But Democritus expanded the idea. He said that all matter—stars, planets, clouds, and people—was made of atoms. Atoms were the smallest things that existed, he said. Nothing can be smaller, and atoms cannot be cut in two. The word *atom* is made up of two Greek words—*a*, which means "not," and *tomé*, which means "cut" or "separate."

The atoms of Democritus were hard and solid and came in

many sizes and weights. They sped about bumping each other and were ageless. The world began, said Democritus, when atoms quite by chance began bumping into one another. When this happened they stuck together, and "matter" was formed. In Democritus's thinking, heavy objects such as rocks and metal were made of heavy atoms. But fire and air were made of much lighter atoms. Atoms of water, he said, were slick and smooth, so they moved over each other easily. But atoms of metal were rough, so they clung together.

Even though the atoms of Democritus came closer to answering the question "What is matter?" than the ideas of those before him, his atoms did not become popular in later years. Their role in the creation of the world did not agree with the Bible. So religious leaders opposed the idea, but it never quite died. And how it changed over time is one of the most remarkable stories in the history of science. Meanwhile, the earth, air,

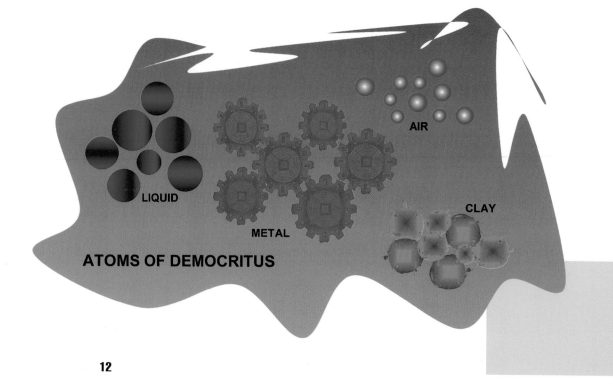

LIQUID

AIR

METAL

CLAY

ATOMS OF DEMOCRITUS

fire, and water theory of Empedocles remained popular and continued to fuel the dreams of the alchemists.

New Elements and New Atoms

The years of the 1600s and 1700s saw a weakening and finally the death of alchemy. They also saw the birth of chemistry as a science. In 1661 the English chemist Robert Boyle published *The Sceptical Chemist*, a book that swept away the earth, air, fire, and water "elements" of the old Greeks. He said that only four elements could not begin to account for all the chemical changes we observe in nature—how our body cells used oxygen, how iron was made into steel, how fudge was made, and so on.

Boyle did many experiments. Melting silver and combining it with other metals, he saw that the silver always remained silver. It never turned into a different substance. Silver was a true *element*, and so was gold, and so was copper, he pointed out. Boyle gave the word *element* the meaning it has today: an element is any substance that cannot be broken down into a simpler substance or built up from a simpler substance. How amazed Boyle would have been had he known that one day atomic scientists would learn to make true elements that we have not yet found in nature! Today we know of 112 elements, only 90 of which occur in nature.

Despite Boyle's work, many who called themselves scientists still clung to the old earth, air, fire, and water notion of "elements." Yet others were stirred up by his findings and looked to the laboratory, not the philosophers, for scientific knowledge.

Democritus, who lived around 450 B.C., taught that all matter—stars, planets, clouds, and people—was made of atoms. Atoms were the smallest things that existed and could not be cut in two. Some atoms were smooth, others rough, some heavy, others light. Liquids were made of smooth atoms that could slide past each other. Solids were made of rough atoms that stuck together.

One such man was the English chemist John Dalton, who lived a century after Boyle's time (from 1766 to 1844). The new idea of elements fascinated him. He knew that many gases were individual elements, such as oxygen and hydrogen. Yet oxygen could join with carbon to make the gas carbon dioxide. Dalton reasoned that the atoms of the different elements must somehow be different from one other, but the atoms of one element must all be the same. But how could he be sure? He could not see atoms to find out how they were shaped. He knew of no way to weigh them. He supposed, as had Democritus, that all atoms were minute globes of matter that could not be broken into smaller pieces.

When the atoms of one element somehow join atoms of

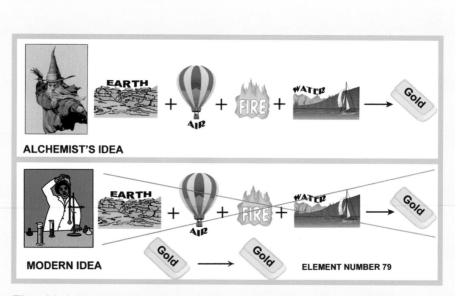

The old alchemists believed that just the right combinations of their four base elements would become gold or silver. In the 1600s the English chemist Robert Boyle did experiments that proved that silver or copper, for example, could be combined with other metals, but the silver or copper always remained the same. They never turned into a different substance. Boyle gave the word element *the meaning it has today.*

another element and produce a new substance, the atoms them-selves do not change. By Dalton's time, it was known that water is made of hydrogen and oxygen. In his chemical picture writing he pictured water as one atom of hydrogen joined to one atom of oxygen. When such a joining took place, said Dalton, a "com-plex atom" was formed. Today we say that a *compound* has been formed. Dalton imagined his simple atoms had spines that allowed them to cling together as complex atoms.

Dalton's idea about atoms of hydrogen joining with atoms of oxygen and forming a new substance, water, was sound even though it was not quite correct. Experiments by a French scien-tist showed something that confused Dalton. When the gas water vapor was formed by igniting hydrogen and oxygen in a tube, the space taken up by the hydrogen was always twice greater than the space taken up by oxygen. That experiment suggested that the chemical formula for a particle of water should be written like this:

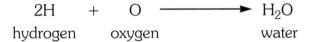

$$2H \quad + \quad O \longrightarrow H_2O$$
hydrogen   oxygen   water

At first Dalton refused to accept the idea, but then the puz-zle was solved by the Italian physicist Amedeo Avogadro. He showed that the smallest unit of a compound such as carbon dioxide was not the atom, but a cluster of atoms. He called such a cluster a *molecule*, which nicely explained Dalton's "complex atoms." So a molecule is the smallest possible amount of any substance that still acts exactly like larger amounts of the sub-stance. One molecule of carbon dioxide consists of one atom of carbon joined to two atoms of oxygen (CO_2). One atom of the element sodium (Na) joined to one atom of the element chlorine, as NaCl or sodium chloride, is the smallest possible piece of salt,

The English chemist John Dalton improved on the idea of atoms by supposing that the atoms of the element oxygen must all be alike but different from the atoms of hydrogen, for instance. He also supposed that atoms were the smallest possible pieces of the chemical elements. Dalton lived from 1766 to 1844.

or a single molecule of salt.

By 1844, when Dalton died, scientists all over the world had come to accept Boyle's idea of elements, Dalton's idea of atoms, and Avogadro's idea of molecules. They were exciting ideas, for they were bringing scientists ever closer to an understanding of what matter was. But there were questions. There always are questions. What force locks the atoms of a molecule together one moment but then a moment later frees the atoms to go their separate ways? Would it ever be possible to measure the size of atoms? Was it possible that atoms might not be solid, unbreakable objects? Although atoms were the building blocks of matter, could atoms themselves be made of even smaller building blocks?

Today we imagine atoms as spheres of different sizes and mass. Different kinds of atoms join and form the many gases, liquids, and solids we find around us. The crystal solid shown here is quartz, which is made up of silica (a combination of the elements silicon and oxygen). Its atoms are locked rigidly together in regularly repeating patterns.

Three

Inside the Atom

Thomson Discovers Electrons

The scene is the famous Cavendish Laboratory at Cambridge University, England. A young physicist named J. J. Thomson (1856 to 1940) has just darkened the room, and he stares at a glass tube about the size of a cucumber. The tube glows with an eerie light. Thomson had pumped nearly all of the air out of the tube, then run electricity through it. The result was a pale glow made by rays that stretched along the tube. What puzzled Thomson was not the stream of rays but that he could bend them with a strong magnet. Try as he did, he could not bend ordinary light rays with a magnet. The rays were called *cathode rays*, the same rays that flow through the tube of a television set and produce a picture. Since cathode rays could be bent by a magnet, then they were not ordinary light rays. What were they? Could they be matter? Could they be pieces of atoms? Thomson experimented with cathode-ray tubes for twenty years.

In 1897 he presented his findings before a meeting of England's most respected scientists, the Royal Society. His listeners could hardly believe what Thomson told them. After all, there were still physicists who did not even believe in atoms. "Cathode rays," Thomson announced, "are particles of negative electricity." They are pieces of electricity, which Thomson called "corpuscles," and are tiny parts of atoms that can be stripped away from the main part of the atom. All atoms, he said, have corpuscles, and all corpuscles are alike. They are the same size and the same weight. With that announcement, the *electron* was born. Thomson's "corpuscles" were given the name electrons by an Irish physicist, George J. Stoney. Thomson's ideas were very convincing and gained him worldwide attention.

The Thomson atom was a blob of something that somehow managed to hold on to the electrons within it. But how? Since an electron had a negative electric charge, Thomson wondered if there might be positively charged particles scattered around along with the electrons. If not, he reasoned, then the atoms of our skin or of the air would be electrically charged, but they are

J. J. Thomson at work in England's Cavendish laboratory at Cambridge University. A British physicist, Thomson won the Nobel prize in 1906 for his discovery of electrons. He lived from 1856 to 1940.

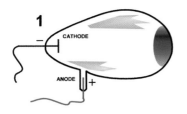

In Thomson's experiments, a glow of green light appeared at the end of a glass vacuum tube when an electric current passed through the tube between the cathode, or negative plate, and the anode, or positive plate.

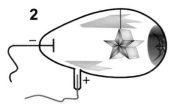

Thomson thought the glow must be caused by some kind of "rays" traveling straight out from the cathode, because an object placed in their path cast a shadow on the end of the tube.

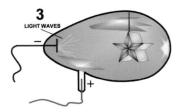

He knew that the "rays" could not be light waves from the heated cathode. If they were, they would spread out in all directions, making most of the tube glow and producing only a blurry shadow behind the object in the tube.

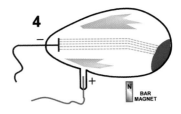

When the "rays" were pulled toward the end of a bar magnet held near the tube, Thomson concluded that they must be made up of a stream of particles, and that each particle carried an electric charge.

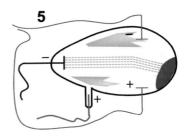

When Thomson passed the stream of particles between two electrically charged plates, the stream was bent away from the plate with the negative charge. This told him that each of the particles must carry a negative charge, since like charges were known to repel each other.

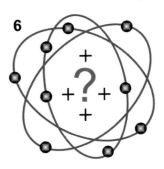

Thomson said that these negatively charged particles (electrons) are part of all atoms. But he felt that some part of an atom must carry positive electrical charges to balance the negative charges and hold the electrons in place.

not. They are electrically neutral. For each negative charge of an atom's electron, Thomson reasoned further, there must be a positive charge carried by some other part of the atom.

If that is so, Thomson asked himself, then where within an atom were these pieces of positive charge? Maybe they were lumped together at the center of the atom. Maybe they were scattered around throughout the atom, like the raisins in a plum pudding. He didn't know, but he guessed that they were scattered around. As it turned out, his guess was wrong. It was one of Thomson's students who would solve the puzzle. Nevertheless, Thomson had managed to open up the atom for other scientists to explore.

How Rutherford Solved the Puzzle

One of Thomson's most brilliant students was a young physicist named Ernest Rutherford. He was from New Zealand and had won honors in just about every subject he studied—mathematics, physics, chemistry, Latin, French, and English literature. His experiments to penetrate to the heart of the atom began in the Cavendish Laboratory around 1896. They also took him to Canada's McGill University and continued for more than twenty years.

In his quest for the atom's heart, Rutherford experimented with a strange element called thorium. At the time only three other elements like it were known. Today we call them *radioactive* elements. The atoms of all three were so big and heavy that they were unstable and tended to break apart. Bits and pieces of them would fly off in all directions as tiny "bullets." By putting a piece of thorium in a container with a tiny window, Rutherford could direct the stream of thorium bullets by aiming the container.

In one experiment to learn more about these bullets, he fired

Ernest Rutherford, a British physicist born in New Zealand, discovered that the atom is mostly empty space and identified the nucleus of the atom. He also discovered alpha, beta, and gamma rays. He lived from 1871 to 1937.

streams of them at a thin sheet of gold foil. He wanted to find out what the atoms of gold might do to the ejected thorium particles. Possibly it would tell him the way the atoms were put together. A special screen behind the gold foil showed him what happened to the thorium bullets on passing through the gold foil.

Rutherford found that most of them simply smashed through the barrier of gold foil and followed a straight-line path to the screen. It was as if nothing at all had been in their way. Some of the bullets, however, were knocked a bit off course and struck out near the edges of the screen. A few were even bounced directly back toward the thorium "gun." Anyone doing this experiment today would find

To explore the inside of atoms of gold, Rutherford aimed a stream of "bullets" from the element thorium at a thin piece of gold foil. A screen that glowed wherever a bullet struck showed that only a few of the bullets had been knocked off course as they passed through the piece of gold foil.

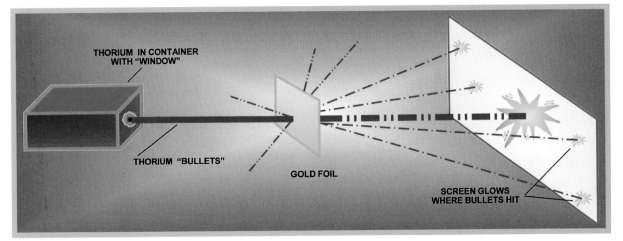

THORIUM IN CONTAINER
WITH "WINDOW"

THORIUM "BULLETS"

GOLD FOIL

SCREEN GLOWS
WHERE BULLETS HIT

that about one out of every ten thousand thorium bullets is knocked off course on passing through the gold foil.

Rutherford's experiments told him two important things about atoms: 1) They must be mostly empty space, since most of the thorium particles passed right through the gold foil barrier. 2) The central core of an atom must be much more massive than the electrons. Otherwise, thorium bullets would not bounce back off the central core. Rutherford wrote that "it was almost as incredible as if you fired a 15-inch shell at a piece of tissue paper and it came back and hit you." Rutherford had discovered the atomic *nucleus*.

The nucleus of an atom contains tiny but massive particles called *protons*. It turned out that a proton is nearly two thousand

1

Thales, about 650 B.C., taught that all matter was composed of water.

2

Heracleitus, about 500 B.C., said that everything was made of fire.

3

Empedocles, about 450 B.C., said the world-stuff was composed of earth, air, fire, and water.

4

Democritus, also around 450 B.C., offered his atomic theory. All things, he said, are made of slick or rough and heavy or light atoms.

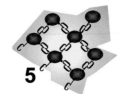

5

Gassendi, about 1630, said that atoms must be hooked together to form solids.

6

Dalton, in 1802, thought that atoms of different substances must differ in size and weight.

times heavier than an electron. As Thomson had suspected, an atom's central core carries a positive charge, which balances an electron's negative charge. To appreciate the fact that an atom is mostly empty space, picture one this way: if the nucleus were the size of a pea, then its nearest electron would be about 150 feet (46 meters) away.

The story of the new "nuclear" atom did not end with Rutherford's discovery of protons, and it has not ended yet! In 1932, some fifteen years after protons were discovered, the English physicist James Chadwick found a second kind of nuclear particle. Called a *neutron*, it has no electric charge. It simply adds mass to an atom and weighs very nearly the same as a proton. Since atoms ordinarily have the same number of protons

7

Lord Kelvin, in 1867, said that atoms were shaped and moved like smoke rings.

8

Thomson, in 1897, pictured the atom as a positively charged globe with electrons inside.

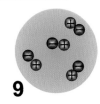

9

Philipp Lenard, in 1903, saw the atom as pairs of positive and negative charges called "dynamids."

10

Nagaoka, in 1904, pictured the atom with a positive nucleus surrounded by rings of electrons.

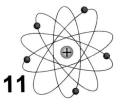

11

Bohr, in 1913, assigned each electron a definite orbit (called the ground state) to move in; but an electron could also jump to a higher orbit or fall to a lower one.

12

Today we view an atom's electrons much as Bohr did, but as being free to move every which way within a given energy level "orbit."

as electrons, they do not carry an electric charge. So electrons, protons, and neutrons are the three basic building blocks of all atoms except hydrogen. Hydrogen is the simplest of all atoms. It has a single electron and a nucleus of a single proton. That's it. The first four simplest atoms are hydrogen, helium, lithium, and beryllium.

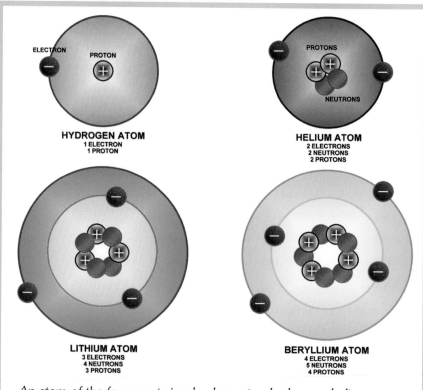

HYDROGEN ATOM
1 ELECTRON
1 PROTON

HELIUM ATOM
2 ELECTRONS
2 NEUTRONS
2 PROTONS

LITHIUM ATOM
3 ELECTRONS
4 NEUTRONS
3 PROTONS

BERYLLIUM ATOM
4 ELECTRONS
5 NEUTRONS
4 PROTONS

An atom of the four most simple elements—hydrogen, helium, lithium, and beryllium. An atom normally has the same number of electrons as protons, so their negative (-) and positive (+) electrical charges are canceled out, and the atom itself is without an electrical charge. Neutrons do not have a charge. They are packed together with protons in the atom's nucleus. The nucleus makes up nearly all of an atom's mass, or weight.

As our story of the atom unfolds, the scene now shifts to France and to the work of physicists Antoine-Henri Becquerel and Marie and Pierre Curie. The Curies had been experimenting with a strange new element that glowed in the dark. It was called uranium. They also discovered other such elements—polonium and radium. They wondered what could possibly cause these elements to glow. And just as baffling, what caused them to give off mysterious rays that could bombard their way through paper and thin sheets of aluminum, and through gold foil like a bullet through a cloud? The answer, Marie Curie thought, must lie hidden deep inside the atoms of these elements.

It was not the Curies but Rutherford and the chemist Frederick Soddy who solved this mystery—not however, before the case of the mysterious rays began to unfold, in 1896.

Four

Case of the Mysterious Rays

Roentgen Discovers X-rays

About the time Rutherford was working with Thomson in the Cavendish Laboratory, in the late 1800s, a startling story appeared on the front pages of newspapers the world over.

A physics professor in Germany had discovered a way to photograph a person's bones! His name was Wilhelm Konrad Roentgen, and the time was December 1895. Like Thomson, Roentgen was experimenting with cathode-ray tubes, but his interest was not in the stream of particles that flowed through the tube. Instead, it was in the bluish glow at the end of the glass tube where the stream of particles struck. Mysterious rays of some sort came from the patch of glowing glass. He found he could get a

By accident, Wilhelm Roentgen discovered that invisible rays of some kind coming from a cathode-ray tube made the chemical coating on a sheet of paper glow. He tried to block the rays by covering the tube with black cardboard, but the rays passed right through the cardboard and made the paper glow whenever electricity was flowing through the tube.

better look at the glowing patch by covering the tube with black cardboard and viewing the glow in the dark. He was startled when he noticed that a piece of paper across the room and coated with certain chemicals glowed each time he turned on the cathode-ray tube. It stopped glowing when he turned off the tube. He was even more surprised to find that the rays from the tube's glowing patch went right through the wall and made the same piece of paper glow when he placed it in the next room! The rays could pass through glass, human flesh, paper, and many other materials. However, they were stopped by a metal key, bone, and certain other hard objects. Because Roentgen did not know what the rays were, he called them *X-rays*.

Wilhelm Konrad Roentgen discovered X-rays in 1895. He was a German physicist who lived from 1845 to 1923.

COATED
PAPER

ANODE

CATHODE

CATHODE
RAY TUBE

Roentgen wrote a report of his discovery and mailed it to several important European scientists the Saturday after Christmas 1895. He also enclosed copies of his X-ray photographs.

Becquerel and Fluorescence

Our story next takes us to France. Nineteen days later, the French physicist Antoine Henri Becquerel learned about Roentgen's X-rays. He was fascinated because he had been studying other substances that glow, or *fluoresce*. They became fluorescent when exposed to *ultraviolet* light, invisible light from the Sun that gives us sunburn. Could it be, he asked himself, that those glowing substances also give off X-rays? In other words, were X-rays a normal part of *all* materials that fluoresce?

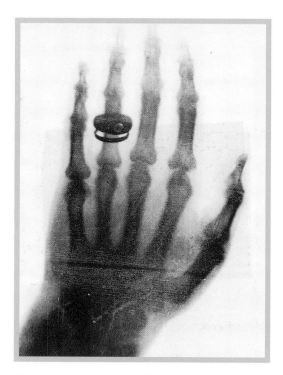

This is one of the first X-ray photographs made by Roentgen. It shows a hand with two rings on one of the fingers.

Becquerel knew that certain minerals in rocks glowed when exposed to ultraviolet light. So he supposed that the glow came from the energy of sunlight. Somehow, fluorescing mineral crystals exposed to light glowed as long as they received ultraviolet light. But as soon as they were removed from the ultraviolet light, they stopped glowing.

In one experiment Becquerel used certain fluorescent crystals that happened to contain some uranium, but also contained the elements sulfur, potassium, and oxygen. He wrapped a photographic plate (film had not yet been invented) in black, light-proof paper. He then sprinkled a few of his crystals on top of the paper and placed the package outside his window in sunlight. He predicted that the sunlight would make the crystals fluoresce and so give off rays that would expose the photographic plate. When he developed the plate, sure enough, he found little gray smudges in just the places where the crystals had sat. Didn't this prove that sunlight somehow energized the crystals to give off rays that passed through the light-proof paper and so exposed the plates?

So it seemed. But then one day when Becquerel was about to do the experiment again, clouds moved in and hid the sun. Discouraged, he placed his carefully wrapped photographic plate in his desk drawer, scattered the crystals on top, and closed the drawer. Surely, he thought, nothing would happen to the plate, or to the crystals, since they were in the dark. There would be no ultraviolet light to make them glow and give off rays. A few days later Becquerel became curious and decided to develop the photographic plate. Imagine his surprise when he found gray smudges on the plate where the uranium crystals had rested. But this time the smudges were even sharper and darker than before!

So light was not needed to make the crystals give off their mysterious rays. They were doing it on their own, and in the dark. He repeated the experiment again and again, each time with the same results. Even if he kept the crystals in the dark for several weeks, they still produced the rays.

Becquerel next decided to test all the fluorescent substances he could find. Although some produced the rays, others did not. Those made of calcium or zinc, for instance, did not. It turned out that only those that contained uranium gave off the rays. And neither ultraviolet light nor visible light was needed to start up the rays. Becquerel next decided to test a sample of pure uranium, not just a rock that contained bits and pieces of uranium. Maybe the rays would be stronger. Although pure uranium was hard to come by in 1896, he managed to get a sample from a chemist friend. As before, he placed the sample on top of the light-proof heavy black paper wrapped around a photographic plate and closed it in his desk drawer. When he later developed the plate he found a very strong black area directly beneath where the uranium had rested. Becquerel then knew beyond doubt that uranium was a storehouse of energy and gave off rays anywhere, anytime. But what *were* the rays?

A number of different kinds of rocks become fluorescent when held under black light, which is ultraviolet or infrared radiation. The rock here contains the mineral calcite, which glows pink. It also contains willemite, which glows greenish-blue.

Five

Atoms That Change by Themselves

The years just before 1900 were exciting ones for physicists. Rutherford and Soddy were off in Canada trying to learn more about Becquerel's rays, as Becquerel himself was trying in France. Roentgen had recently discovered X-rays, which everyone was interested in but knew nothing about. Also in France, Marie Curie was trying to solve the mystery of the rays given off by uranium.

Eight Tons of Pitchblende

The year was 1897, and the question Marie Curie asked herself was: how does a piece of pure uranium get the energy that causes it to give off Becquerel's mystery rays? She began by testing every different kind of metal known to her to see if it gave off the rays. In only one case did she find one, other than pure uranium.

It was the mineral ore pitchblende. But she had expected pitchblende to give off the rays because it contained uranium. Pure uranium was obtained by breaking down the pitchblende rock and separating out the pieces of uranium metal. What she did not expect was that the pitchblende's rays were even stronger than those of pure uranium.

A pound of pitchblende gave off stronger rays than a pound of pure uranium. How could that be? The pound of pitchblende contained rock and other matter along with only a small amount of uranium. The only answer Marie could come up with was that the pitchblende contained an additional substance that also gave off the rays. Here was a new idea to test.

Marie and her physicist husband, Pierre, spent several months breaking down huge amounts of pitchblende into its several different elements. Finally, in July 1898, they found a new and unknown element. It appeared only as a smudge at the bottom of a test tube. It was not uranium but it gave off the rays. Marie later named the rays *radioactivity*, and she named the new element polonium, after Poland, where she was born. All elements that gave off the rays came to be called radioactive elements.

The following December the Curies discovered still another new element that also gave off the rays. This one was about two million times more radioactive than uranium! They named the element radium. Their radium prize was hard won. To get only less than an ounce of it, they had to break down eight tons of pitchblende.

New Atoms from Old Atoms

By the early 1900s physicists all over the world had become fascinated by this newly discovered form of energy called "radioactivity." In England, Sir William Crookes, inventor of the cathode-ray tube, discovered a puzzling new substance while experimenting with uranium. It was so much like uranium that he called it uranium-X. In Canada, Rutherford and Soddy were experimenting with the radioactive element thorium. They, too, discovered a puzzling new substance. It was so much like thorium that they called it thorium-X.

Could thorium somehow change into thorium-X, and uranium into uranium-X? The Curies thought no. The radium they had produced had been in storage for several months and showed no signs of changing. Rutherford and Soddy thought yes. They were convinced that certain kinds of atoms broke down all

This illustration shows Marie and Pierre Curie in their laboratory in Paris, where they discovered the elements polonium and radium. Marie Curie became the world's most famous woman scientist and the first person to win two Nobel Prizes— science's highest award.

by themselves and changed into atoms of a different element, thorium into thorium-X, for example. As the atoms broke down, they released energy in the form of radiation.

If true, then what exactly was this "radiation"? Could it be pieces of big and unstable atoms? To find out, Rutherford designed a new experiment. As with his thorium "bullets" experiment that led to the discovery of protons, he built a box with a small window. A radioactive element placed in the box would shoot radiation out through the window at a target. In this case, the "target" was the space between two magnets. In this way, Rutherford experimented with the radiation from the radioactive elements known to him. What did he find?

Identifying the "Rays"

Were the rays bits and pieces of matter, like Thomson's electrons? Or were they waves of energy, like light? What else *could* they be? Some of the radiation that Rutherford fired through a magnetic field was bent sharply in one direction. That radiation turned out to be high-speed electrons. He gave them the confusing name of *beta particles*.

The radiation also contained other particles. These were bent in the opposite direction as they passed through the magnetic field, but they were bent less sharply. This told Rutherford that the particles must carry a positive electric charge since they were bent in a direction opposite to that of the negatively charged electrons. And the fact that a stream of these other particles was

Rays given off by radioactive elements are bent when they pass through a magnetic field—the space through which the opposite poles of two magnets attract each other. Beta particles (electrons) are bent sharply in one direction. Alpha particles (clumps of two protons and two neutrons each) are bent only a little, in the opposite direction from beta particles. Gamma rays are not bent at all.

bent less sharply than the stream of electrons suggested that the particles were heavier than electrons. Rutherford called the positively charged radiation *alpha particles*. They turned out to be four times heavier than a proton.

Could that mean that each alpha particle was a clump of four protons? If so, then an alpha particle should have a positive electric charge four times stronger than a single proton. But it didn't. Each alpha particle had a positive charge equal to that of only two protons. Why, if it weighed as much as four protons?

Solving the Alpha Particle Puzzle

Rutherford's nuclear atom was being a problem, but just the kind of problem physicists love. The thinking went something like this: was it possible that two of the four protons of an alpha particle each had an attached electron? If so, the negative charge of each electron would cancel out the positive charge of its proton partner. The alpha particle would then be left with the two positive

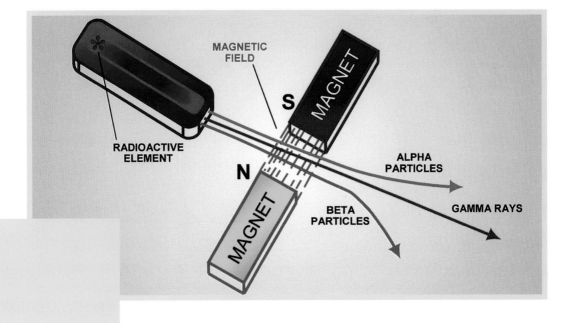

charges of the other two protons without electron partners. The idea sounded very convincing, so convincing that the new alpha particle model hung around for thirty years. But Rutherford wasn't happy with the model. He felt that a six-piece clump of radiation was too clumsy to be true. It wasn't until 1932 that Chadwick discovered the second basic nuclear particle, the *neutron*. Almost immediately, the German physicist Werner Karl Heisenberg realized the importance of Chadwick's discovery. An alpha particle, he said, did not have two electrons stuck to two of its four protons. Instead, each alpha particle was made of two protons and two neutrons. That explained why an alpha particle

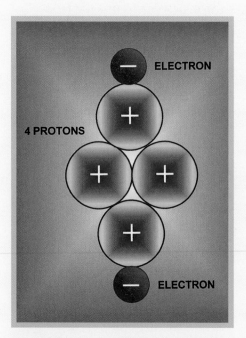

For thirty years scientists thought an alpha particle was made up of four protons and two electrons. The electrons' negative charges were thought to cancel out the positive charges of the two protons.

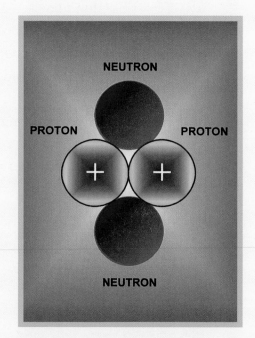

When James Chadwick discovered neutrons—particles with the mass of a proton but without an electrical charge—Werner Karl Heisenberg realized that an alpha particle is made up of two protons and two neutrons.

had the positive charge of two protons but the mass of four protons. An alpha particle is, in fact, the same as a helium nucleus.

During his experiments with beta particles and alpha particles shot through a magnetic field, Rutherford found a third type of radiation. Recall that beta particles were bent sharply in one direction and alpha particles less sharply in the opposite direction. A third type of radiation was not affected by the magnetic field. The radiation sailed straight through. This led Rutherford to suppose the radiation was not made of particles, and he was right. Instead, it was energy traveling as waves, extremely high energy that he named *gamma rays*. Gamma rays are high-energy X-rays and even more destructive than ordinary X-rays. Only a few radioactive elements emit gamma radiation. The radioactive element radium that Marie Curie experimented with is one. She died in July 1934 at the age of sixty-seven. The cause of her death was cancer of the blood, or leukemia. The most likely cause was her exposure to radium during her years of scientific work. That work won her two Nobel prizes, science's highest award.

From the time of Democritus to around 1850 the atom was more an idea than an actual piece of the world. Then, over the next eighty years, atoms became very real things. You could bump into them. You could steer them around with magnets. You could take pictures with them. Physicists and chemists learned how atoms of one element could change into atoms of a different element. They also began to learn how to take atoms apart and package them for use.

Six

The Maverick Atoms

As Alfred Romer said, "an idea in science lasts only as long as it is useful." As we began to see in the previous chapter, a lot of very useful ideas about the atom were coming thick and fast in the late 1800s and early 1900s. And they kept coming.

The Puzzling Case of Heavy Neon

Up until the early 1900s hardly a scientist alive questioned the idea that all atoms of gold, or all atoms of lead, or all atoms of oxygen were exactly like each other. It is when someone questions the truth of a long-held belief in science that things can get pretty exciting. In 1912 two such scientists did just that. One is

J. J. Thomson and Francis Aston sent a stream of charged neon atoms through a magnetic field. The field pulled the atoms downward onto a photographic glass plate (film). The developed film had two gray spots instead of one. This showed that some of the atoms were lighter weight than the others, because they were pulled downward in a sharper curve.

our old friend J. J. Thomson, and the other was the English chemist Francis W. Aston. They were working with neon, the gas that you see glowing in those glass-tube advertising signs. Neon has a somewhat heavy nucleus of ten protons and ten neutrons.

They shot streams of neon nuclei through a magnetic field so that the nuclei struck an unexposed photographic plate. Since the electrons surrounding each nucleus had been stripped away, each nucleus was left with a positive electric charge. That meant that the stream of nuclei would be bent along a curved path as it passed through the magnetic field. If each neon nucleus were exactly alike, then all the nuclei would follow the same curved path and strike the photographic plate in one small neat bull's-eye patch. But that didn't happen. Instead, there were two patches. Some of the nuclei had been curved more strongly than

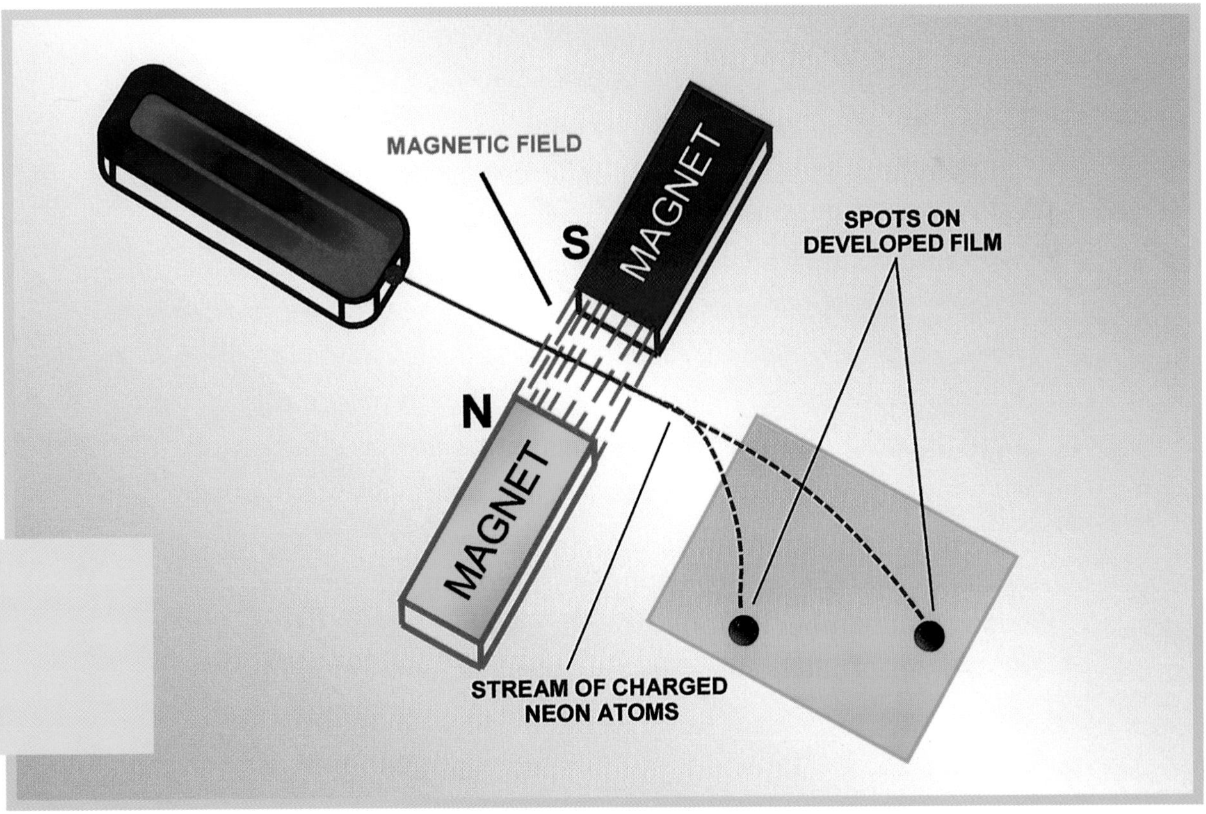

MAGNETIC FIELD

MAGNET

S

N

MAGNET

SPOTS ON
DEVELOPED FILM

STREAM OF CHARGED
NEON ATOMS

others. Why? The most likely answer was that there were two different kinds of neon atoms, some heavier than the others. The stream that was curved sharply contained lighter-weight atoms. Like a charging elephant, the heavier atoms were harder to curve off course. It was left to Soddy to name such maverick atoms of the same element *isotopes*, from the Greek words *iso* and *topos*, meaning "same place." Isotopes of an element turned out to have the same number of protons but a different number of neutrons. That made them behave exactly alike chemically but gave them different weights.

How to Weigh an Atom

The Thomson and Aston experiments with neon gave other scientists a way to measure the weight of atoms. How fascinated Democritus and Dalton would have been by all this. The idea was really quite simple. By comparing the amount of curving of a heavier isotope with the amount of curving of a lighter isotope, you could say that one of the isotopes was two, or fifteen, or twenty-five, or whatever times heavier than the other.

Although some elements have only two isotopes, others have more. The table on page 41 shows five of the seven isotopes of carbon. Notice that each isotope has exactly the same number of protons and electrons. That makes the atom as a whole electrically neutral, since each negatively charged electron cancels out the positive electrical charge of each proton. Again, it is the differing number of neutrons that makes one carbon isotope heavier or lighter than another. We abbreviate carbon with the capital letter C, and we show which isotope of carbon we are talking about by adding its number of protons and neutrons and writing that number in front of the C, like this: ^{10}C or ^{11}C, or ^{12}C, and so on. That number is called the atom's *atomic weight*.

Atoms also have what is called their *atomic number*. The atomic number is simply the number of protons in the nucleus. So ^{14}C has an atomic weight of 14 and an atomic number of 6.

Isotope of carbon	Number of protons	Number of electrons	Number of neutrons
Carbon-10	6	6	4
Carbon-11	6	6	5
Carbon-12	6	6	6
Carbon-13	6	6	7
Carbon-14	6	6	8

Finally, the puzzle of uranium-X and thorium-X could be solved. And it was to the credit of Soddy. While experimenting with the radioactive element thorium, he discovered that thorium had at least two different kinds of atoms, or isotopes. One had an atomic weight of 232 and the other had a weight of 228. They were simply two different forms of the same element. That very important idea had been overlooked by Thomson and Aston. Today we know that thorium has at least twelve isotopes.

The idea of isotopes cleared up another problem that had been troubling scientists in different countries. For example, a scientist trying to find the atomic weight of lead mined in Germany would come up with one weight. Then a scientist working with a lead sample from England would come up with a different atomic weight, and so on in different parts of the world. The puzzle was solved when they realized that lead samples taken from different places might well contain different mixtures of lead isotopes. Everything seemed to be falling into place nicely. Well, almost everything.

There always are problems. To find this one we have to back

up a few years to the time when Rutherford had discovered the "nuclear" atom. His model of the atom had a nucleus with a positive electric charge surrounded by electrons with a negative electric charge. The question that *had* to be asked was a bothersome one. What kept the electrons in orbit around the nucleus? Why didn't the positive charge of the nucleus swallow up the electrons by pulling them right down into the nucleus? The only explanation Rutherford could come up with was that electrons stayed in orbit about the nucleus because they were like the planets orbiting about the Sun-nucleus. Good try, but the explanation didn't win many believers. After all, it had long been known that it is gravity that keeps the planets in orbit about the Sun. Atoms involved electrical forces, not gravity. And the nature of electrical forces means a kind of "friction" on an orbiting charged particle (such as an electron) that causes it to lose energy and spiral into the nucleus. It was left to another modeler of the atom, the brilliant physicist Neils Bohr, to solve that puzzle. We will return to Bohr in the final chapter, when the atom becomes really interesting.

How to Make an Isotope

Soddy had discovered the neutron. He also had explained isotopes. He and Rutherford had done a lot of experiments with thorium "bullets," or alpha particles. The stage was now set for another important face of the atom to be unmasked. The year is 1934 and the place the laboratory of Marie Curie's daughter Irène and her husband, Frédéric Joliot-Curie. Recall that an alpha particle is a clump of atomic nucleus that contains two protons and two neutrons. They wanted to find out what would happen if they fired a stream of alpha particles at a piece of aluminum. So they set up a polonium "bullet gun" and began firing away.

The atoms of aluminum in your kitchen "tin" cup have an

atomic weight of 27. Each atom has 13 protons and 14 neutrons. When the Joliot-Curies fired alpha particles at the aluminum, they found that the aluminum splattered off neutrons and electrons. They thought that when they stopped the alpha particle bombardment, the aluminum would quiet down. It didn't. It kept splattering off electrons—and only electrons.

The Joliot-Curies had made the aluminum radioactive. Each aluminum atom struck by an alpha particle gained 2 protons. That changed the aluminum into a different element, the element phosphorus. Ordinary phosphorus has 15 protons and 16 neutrons. The phosphorus produced by the Curies was a radioactive isotope, the first artificial isotope knowingly made. Since the Curies' time, physicists have made about a thousand different artificial isotopes.

If physicists can change one element into another in the laboratory, does the same thing happen in nature? The answer is yes, and that's just what radioactivity is all about. Let's take uranium as an example. We'll start with ^{238}U, the most common isotope. Each uranium atom of this isotope has 146 neutrons and 92 protons in the nucleus ($146 + 92 = 238$). Since a uranium atom has a whopping big nucleus, the atom is unstable. All by itself the atom shoots off an alpha particle, or a clump of 2 protons and 2 neutrons. So the ^{238}U atom loses 2 protons and 2 neutrons. As the table on page 44 shows, the loss of an alpha particle changes the ^{238}U atom into an atom of the radioactive element ^{234}Th. This isotope of thorium has 144 neutrons and 90 protons. Now, because the ^{234}Th atom also is unstable, it shoots off a beta particle, or an electron. This changes the thorium atom into an atom of a still different radioactive element, one called protactinium. After a total of thirteen such changes of one radioactive element into another, the final product is the lead isotope ^{206}Pb, with 124 neutrons and 82 protons. Because this isotope of lead has a stable nucleus, it is not radioactive, and the ^{206}Pb remains ^{206}Pb forever.

Kind of atom	Atomic weight	Number of neutrons	Number of protons	Particle lost
uranium	238	146	92	alpha
thorium	234	144	90	beta
protactinium (and so on)	234	143	91	beta
lead	206	124	82	not radioactive

Atoms are wonderfully predictable things. Because we have learned how one radioactive isotope changes into a different isotope, we can put atoms to work for us in many different ways. Telling geologic time is one way.

Seven

Using Atoms to Measure Time

Radioactive Clocks and How They Tick

We can use atoms to measure the age of a rock formation or an old bone that is millions of years old. To understand how such long periods—called *geologic time*—can be measured, you have to know what is meant by the half-life of a radioactive element.

The half-life of a radioactive element is the time it takes for half the atoms in a piece of rock or bone we want to date to change into atoms of another element. Here's how it works: Imagine that you have a large box with 12,800 marbles in it. You want to find out how old the marbles are. Imagine also that the black marbles "age" by turning red. In one year, half of the black marbles turn red. So we say that the black marbles have a half-life of one year. At the end of the first year, there would be 6,400 black marbles and 6,400 red ones. At the end of the second year, half of the remaining 6,400 black marbles turn red. There are now 3,200 black and 9,600 red marbles. This process would go on and on until our clock ran down, when we ran out of black marbles.

Suppose that you counted 800 black marbles and 12,000 red ones. A little simple arithmetic would tell you that the box of marbles was four years old. Very simply, that is how a radioactive clock works. The scientist measures the difference between the number of unchanged atoms of a radioactive element and the number of new isotopes that have formed. Nothing seems to affect the half-life of any radioactive element, neither heat nor pressure nor anything else we know of.

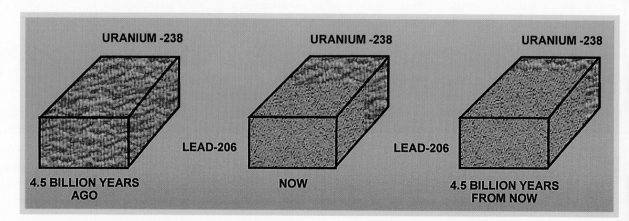

The half-life of a radioactive isotope is the time it takes for half the atoms in a sample of the substance to decay, or change into atoms of another isotope. For example, in 4.5 billion years, half the uranium-238 atoms in a piece of rock change into atoms of lead-206. After 4.5 billion years more, half the remaining uranium-238 atoms will decay, and so on. Other radioactive isotopes have half-lives measured in thousands of years, in days, in hours, or in fractions of a second.

Different radioactive elements have different half-lives. So scientists have a closet full of radioactive clocks to measure very long periods of geologic time or much shorter periods of time. Here are four such clocks.

This radio-active element	changes into this radioactive element	and has a half-life of
uranium-238	lead-206	4,510 million years
potassium-40	argon-40, calcium-40	1,350 million years
rubidium-87	strontium-87	50,000 years
carbon-14	nitrogen-14	5,730 years

Mosquitoes and Old Bones

Let's look into the carbon-14 radioactive clock to find out how biologists use it to find the age of once-living matter. The isotope ^{14}C is present in the atmosphere in small amounts, and the amount seems always to stay just about the same. Living plants take in the isotope. Animals then eat the plants, and so they also take in the isotope. From the moment a plant or animal dies, it stops taking in carbon. Also at that moment the amount of ^{14}C in its body begins to get smaller. This happens as the radioactive carbon-14 atoms begin to change into atoms of nitrogen.

Suppose that you are an archaeologist and dig up the remains of a campsite that you suspect is several thousands of years old. Your problem is to measure its age by using the carbon radioactive clock. First you select a piece of old wood about the weight of a cigar from the campsite. Next you cut a piece exactly the same weight from a living tree. You have an instrument called a radiation counter and hold it next to the new piece of wood. The instrument begins giving off beeps, each beep signaling one ^{14}C atom changing into the nitrogen isotope ^{14}N. You carefully time the beeps for exactly one minute and count twenty beeps. Next you do the same thing with the old piece of wood, but you count only ten beeps a minute. This would tell you that the old piece of wood had lost half the number of ^{14}C atoms it had when it was part of a living tree. Since it takes 5,760 years for half the ^{14}C atoms in a dead animal or plant to change into ^{14}N, this tells you that the old piece of wood was cut from a tree about 5,700 years ago.

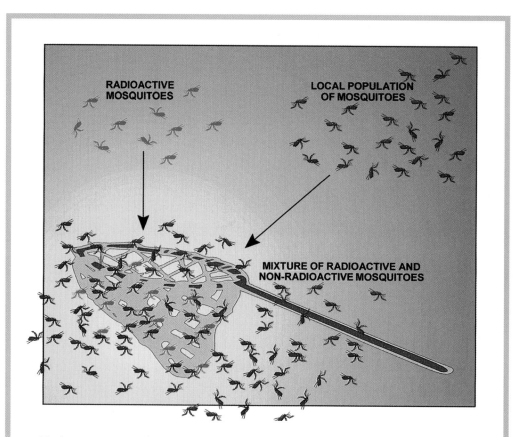

Biologists can use the radioactive isotope strontium-90 to help them count populations of insects and trace their travels. Strontium-90 is fed to a group of mosquitoes in the laboratory (red) to make the insects radioactive. Because they are radioactive, they can be detected by a radiation counter. One hundred of the radioactive insects are released in an area where biologists want to count the local mosquito population. The radioactive mosquitoes quickly mix in with the non-radioactive members (blue) of the population. When a sample of 100 mosquitoes are captured in the area and then tested for radioactivity, only 10 are radioactive. Since the radioactive mosquitoes are mixed evenly within the entire population, there must be about 10 times 100, or 1,000, mosquitoes in the local population.

Other Uses for Radioactive Isotopes

Thomson, Rutherford, the Curies, and other pioneer explorers of the radioactive atom would be surprised indeed by how the atom has been put to work in medicine and science. For example, radiation from the artificial radioactive isotope of cobalt 60 (^{60}Co) can be aimed with great precision into the body of a cancer patient. The radiation then homes in on the cancer cells and kills them.

Less powerful isotopes can be injected into a patient's blood. As they are carried along or trapped by certain tissue or organs, they give off signals that tell doctors about the patient's blood circulation. Called *tracer isotopes*, they can reveal how plants use certain nutrients and how well our bodies use what we eat. Those early experimenters with radioactive isotopes never imagined what an important new tool they were preparing for today's medical scientists.

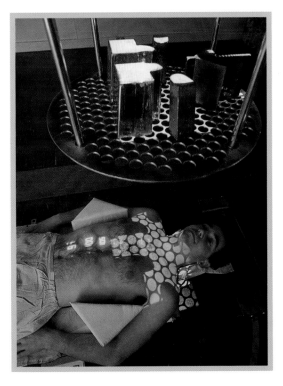

Radiation is commonly used to treat people with cancer. The discs of light shining onto the patient's chest mark target areas that are to receive X-ray radiation. The radiation will kill the cancer cells.

49

Eight

Splitting the Atom for Energy

The Alchemists' Dream Come True

The year Marie Curie died (1934), a thirty-three-year-old Italian scientist named Enrico Fermi was experimenting with uranium. It was then the heaviest element known. Fermi wondered if he could make a uranium atom still heavier by shooting neutrons into its nucleus. He became the first to investigate in detail the effects of shooting neutrons into big atoms. But before finding out how he managed to do it—and the earth-shaking results—a word about the old alchemists. Or rather the new alchemists.

Recall that the old alchemists dreamed of changing copper into gold, and lead into silver, by mixing the copper or lead with just the "right" substance. By the early 1930s a number of

Enrico Fermi, the Italian-American physicist, changed uranium atoms into a new radioactive element by bombarding uranium with slow neutrons. His work on the Manhattan Project during World War II led to the construction of the atomic bomb. In 1938 Fermi was awarded the Nobel Prize in physics for his work. He lived from 1901 to 1954.

scientists were experimenting with isotopes made in the laboratory. For example, if a neutron bullet struck the nucleus of an atom of oxygen-18, the atom became the isotope oxygen-19. The oxygen-19 atom then gave off an electron and changed into a different element, the isotope fluorine-19.

A few years after Fermi began to experiment with uranium, an American physicist managed to turn mercury into gold. The alchemists' dream come true! But he didn't perform his modern magic by using any of the "secret mixtures" of the old alchemists. Instead he bombarded mercury atoms with neutrons and so changed them into atoms of gold. But the few million atoms of

gold he produced couldn't begin to pay the large bill of performing the magic. It is much cheaper to dig gold out of the ground. Now back to Fermi.

The Atom Splitters

The task Fermi set for himself was to change ordinary uranium (^{238}U) into heavier atoms. His plan was to bombard the uranium with neutrons. But the neutrons would have to be slowed down. Ordinarily, neutrons travel too fast to stay close enough to another atom's nucleus to be absorbed and become part of the nucleus. To slow down his neutron bullets, Fermi first shot them through water, or through wax. This slowed them down to just about the speed of air molecules.

As he expected, his slow neutrons were captured and changed the uranium atoms into atoms of a different element. In fact, two different elements, but what those elements were baffled Fermi. And there matters rested.

We now move from Italy to Germany. It is four years later, 1938. The German physicist Otto Hahn thought he had an explanation for Fermi's two mysterious substances. But the idea seemed so wild that Hahn would not even talk about it in public. He did discuss it with his co-worker, the Austrian physicist Lise Meitner. Hahn concluded that the two new substances were isotopes of the element barium. What made the idea at first seem wild was that barium is only half the atomic weight of uranium. There was only one way that two lightweight atoms could be produced from one heavyweight atom of uranium. You guessed it! The uranium nucleus was split in two by the neutron bullet.

Here was a source of energy beyond the wildest dream. But it also was the source of a weapon destructive beyond the imagination—the nuclear bomb.

Lise Meitner, Austrian-born physicist, worked with Otto Hahn and her nephew Otto Frisch. Their work made them realize that they had managed to bring about the fission of uranium atoms. She refused to engage in the work that led to the atomic bomb. Meitner lived from 1878 to 1968.

The Atom Makers

By around 1940 physicists felt pretty much at home with the atom. At that time uranium was the heaviest element known. With 92 protons in its nucleus, it was element number 92. Hydrogen was element number 1, with its single proton. Helium was element number 2, with 2 protons, and so on up to uranium. Could there be still heavier elements? Recall that Fermi had tried to make element number 93 by firing slow neutrons into the uranium nucleus. Although he finally succeeded, he didn't realize his success at the time. In 1940 two American scientists identified element 93. It was given the name neptunium after the planet Neptune. With an atomic weight of 237, the element has a half-life of only a little over 2 million years. That is only a small fraction of uranium's half-life of 4,510 million years.

Later that year radioactive element 94 was made and

named plutonium after the planet Pluto. With an atomic weight of 244, it has a half-life of 24,400 years. Plutonium is an extremely dangerous poison. Over the next ten years University of California scientists, working under Glenn T. Seaborg, made a half dozen more elements. They include the elements americium, element number 95; curium, number 96; berkelium, number 97; californium, number 98; einsteinium, number 99; and fermium, number 100. So far, elements up to number 110 have been made in the laboratory. The heavier the atom, the harder it is to make and the shorter its half-life. Element 102, nobelium, with an atomic weight of 259, has a half-life of only three minutes. One isotope of the element radon, called actinon, has a half-life of only 33.92 seconds. The heaviest atoms made in the laboratory have half-lives measured only in fractions of a second. For that reason they are not found in nature. The heaviest element found in nature is uranium.

The Birth of a Nuclear Bomb

By 1940 Europe was being rocked by World War II, a time when bombs were very important. And the bigger the bomb, the better. Hahn had managed to split atoms of uranium into atoms of barium and krypton. The process, called atomic *fission*, produced a little burst of energy. Physicists soon realized that if the fission process could be kept going long enough in a large enough lump of uranium, a big burst of energy would be produced. In short, a very powerful bomb!

In 1938 Fermi left Italy and moved to the United States. He also was awarded the famous Nobel prize in physics. It was during his work at the University of Chicago that he figured out how to get a pile of uranium atoms to start splitting under control, and keep on splitting, and keep on producing energy that didn't blow him up. Fermi's reasoning went like this: whenever a uranium atom was split, it gave off one or more neutrons, plus energy. Why couldn't those expelled neutrons become bullets to split other uranium atoms in a big pile of uranium? And why couldn't *those* freed neutron bullets split still other uranium atoms? In short, why couldn't a *chain* of reactions be triggered and so produce a steady flow of useful energy?

It was then 1942. The United States had become involved in World War II in Africa, and the government was very interested in Fermi's work. It was so interested that it secretly paid the bills for Fermi's project. It was clear that success in producing a controlled chain reaction could also produce a very uncontrolled one—a superbomb.

Fermi set to work on his controlled chain reaction, but there were two problems. First, there wasn't very much uranium around. At that time uranium wasn't a very useful element. Second, a certain isotope of uranium was needed to keep a strong chain reaction going. While ^{238}U was the most common isotope, it couldn't keep a chain reaction going. The isotope ^{235}U was needed, and that was even harder to come by. Of every 100 atoms of ordinary uranium that come out of a mine, 99 are ^{238}U. Only 1 out of the 100 is an atom of ^{235}U. The digging was hard.

In 1942 Fermi managed to get enough ^{238}U to make a mound-shaped pile about 24 feet (7.3 meters) across. That was large enough to contain enough ^{235}U atoms to keep a chain reaction going, if only a weak one. Fermi's nuclear "furnace" didn't give off very much heat, but it worked. It kept going and proved that the atom could be harnessed as a powerful new source of energy—for peace as well as for war.

Over the next three years, in a huge effort called the Manhattan Project, the United States government experimented, tested, and finally made nuclear bombs. Its test site was Alamogordo, New Mexico. The first bomb, a uranium-235 bomb, was exploded over Hiroshima, Japan, on August 4, 1945. The second, a plutonium-239 bomb, was exploded over the Japanese city of Nagasaki three days later.

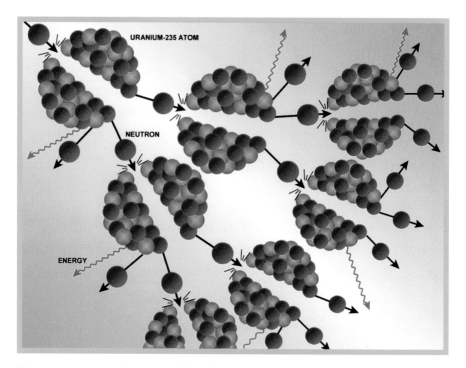

Fission reactions split atoms. The fission chain reaction in an atomic pile begins when a neutron from a decaying atom of uranium-235 (top left) strikes a uranium-235 atom. The struck atom splits into two nearly equal parts. As it does, it gives off 3 neutrons plus some energy. While one of the neutrons is lost, the two others strike two more uranium-235 atoms. Each of those atoms is split into two nearly equal parts. Once again, one neutron from each of the two split atoms is lost while the other two target two more uranium-235 atoms. As before, energy is given off by each split atom. In that way, a chain reaction is set off with the release of huge amounts of energy. The energy can be used to produce electricity or to explode a bomb.

Taming a Chain Reaction

What is the difference between a nuclear bomb and a tame chain reaction in a uranium pile? Very simply, the speed with which the uranium atoms split. During a bomb explosion the freed neutron bullets fly uncontrolled throughout a small lump of relatively pure ^{235}U. The chain reaction they produce goes so fast that all the energy is released in an instant as an explosion.

In a nuclear reactor, the ^{235}U fuel is much less pure and so takes up a lot more space. The freed neutron bullets, then, are less likely to hit other ^{235}U nuclei than in the smaller bomb-lump. Further, many of the freed neutron bullets bump into and are absorbed by shields of graphite rods. In that way the chain reaction is kept slow and under control. Instead of being released in a big bang, the energy trickles out only as fast as a movable graphite shield lets it. Pull out one of the rod shields and the chain reaction speeds up a bit. Pull out another and the reaction goes still faster, with the release of more energy. To produce a chain reaction, your pile of uranium, or plutonium, must be big enough. The smallest amount needed is called the *critical mass*.

From Fission to Fusion

In the 1950s explorers of the atom released a second and much more powerful kind of nuclear energy. A uranium pile releases energy when its slow freed neutrons split apart the nuclei of very heavy atoms. The more powerful kind of nuclear energy came from fusing, or joining, the nuclei of very light atoms. When two such light nuclei are fused, much more energy is released than when a heavy nucleus of uranium is split.

Sunshine comes from just such fusion reactions deep within the Sun's hot core region. There hydrogen nuclei, or single protons, collide, fuse, and eventually build up the nuclei of helium atoms. Actually, it is not ordinary hydrogen that fuses into helium, but two hydrogen isotopes called "heavy hydrogen." One is deuterium, which has a nucleus of one proton and one neutron. The other isotope is tritium, which has a nucleus of one proton and two neutrons. When the nuclei of these two isotopes fuse, they produce one helium nucleus of two protons and two neutrons. At the same time energy is given off, more energy than results from splitting a uranium atom.

Smashing atomic nuclei together to make them fuse is much harder than splitting an atom. The reason is not hard to understand. When you shoot a slow neutron into the nucleus of a uranium atom, the neutron is not pushed away by the positively charged nucleus because the neutron doesn't have any electric charge. It is electrically neutral. Not so when you smash deuterium and tritium together to make them fuse as a helium nucleus. Because the proton in the deuterium particle carries a positive charge, the tritium particle, which also caries a positive charge, tries to push the deuterium away. To get the two to fuse, they must be made to collide with enough force to overcome the mutual pushing away from each other. And that takes lots of energy, which means very high temperatures—millions of degrees!

Where do we find that much energy? We can use a uranium fission bomb to heat up the hydrogen. So a hydrogen fusion bomb of deuterium and tritium is set off by a uranium fission bomb acting as the trigger. When the uranium bomb goes off, it causes the hydrogen nuclei to fuse and release a tremendous amount of energy. It all happens so fast that you see both events as a single explosion.

Physicists have learned to slow down the energy release in nuclear fission reactors and so harness the atom to light and heat our homes. They are now trying to do the same thing with nuclear fusion reactors. What this amounts to is making a miniature Sun in the laboratory. Although they have managed to trigger a safe fusion reaction, they cannot keep one going for more than a small fraction of a second. It is like trying to light a match in a hurricane. Each time you strike one it flashes but then goes out just as quickly. If they could get a reaction to keep going for even one full second, they could then keep it going.

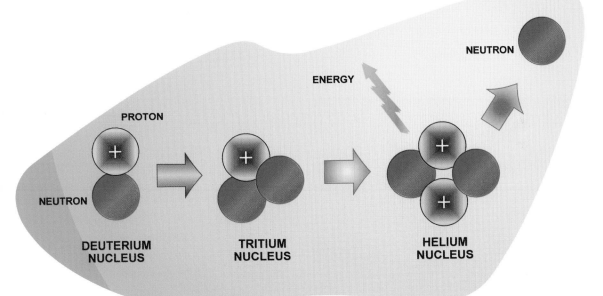

Fusion reactions combine atoms. Intense heat makes the nuclei of heavy hydrogen atoms (deuterium and tritium) fuse, or combine. When they do, they form the nucleus of a helium atom and give off one neutron plus energy. Fusion reactions release more energy than the fission of uranium does. The Sun's huge amount of energy is produced by fusion reactions taking place deep in the Sun's center.

Fusion reactors have two major advantages over fission reactors. At least they would if we could get one going. There is plenty of heavy hydrogen fuel for a fusion reactor. The oceans contain lots of it in the form of deuterium, enough to last for millions of years. There is not nearly so much uranium around. The second major advantage of fusion reactors—if they can ever be perfected—is that they would leave much less dangerous radioactive "ash."

Fission reactors produce many radioactive wastes, and disposing of those wastes has become a serious, and unsolved, problem. Not only do they pose major health threats, but their long half-lives will cause them to linger for tens and hundreds of thousands of years. How do we dispose of those deadly wastes? Or *can* we dispose of them?

There are almost 450 nuclear power plants in the world like this one in Stedman, Missouri, and more are in the making, especially in China. Each plant produces radioactive wastes that must be buried or contained in some way for years, centuries, or thousands of years. Added to the radioactive wastes from such plants are the more dangerous wastes from nuclear bombs and warheads, produced mostly by the United States and Russia.

Nine

Pollution from the Atom

Radioactive Wastes

To this day, no one knows of a safe way to dispose of the radioactive wastes from Fermi's 1942 work at the University of Chicago. It is still radioactive, buried under a foot of concrete and two feet of dirt on an Illinois hillside. And no one knows what to do with the hundreds of thousands of tons of radioactive wastes that have piled up since Fermi's time.

What is radioactive "waste"? It comes in many forms: used uranium fuel whose atoms have been split and can no longer produce useful energy but remain dangerously radioactive. And there are filters and pieces of equipment that have been contaminated with radioactivity and must be discarded. And there are the gloves and other radioactive clothing of workers in nuclear power plants and government nuclear weapons plants. And there is the widespread contamination left in the environment when a test bomb is exploded in the air, underground, or underwater. And there are the millions upon millions of tons of waste material left over from the mining and processing of uranium. Such wastes contain hundreds of different kinds of "hot" radioactive isotopes, which give off several types of radiation, as shown below.

At the close of 1996 there were 436 nuclear reactors in the world, and dozens more are now being planned, many in China. There also were many military bases where nuclear warheads were made and stored. All leak radiation into every part of the

TYPE OF RADIATION	WHAT STOPS IT
Alpha particles	Each alpha particle is a cluster of two protons and two neutrons. These clusters are relatively heavy and slow-moving. The skin stops them, but we can inhale alpha-emitting substances, or swallow them as part of food. They can then emit alpha particles that do damage inside the body.
Beta particles	Each beta particle is an electron. Electrons are more penetrating than alpha particles and can pass through flesh or several inches of wood.
Gamma rays	These are waves of energy and are the most damaging. Gamma rays can pass through several inches of steel or several feet of concrete. They are highly destructive to living tissue.
Cosmic rays	Radiation from outer space. Cosmic rays contain all three kinds of the radiation listed above.
X-rays	Produced by medical and dental X-ray machines, these rays can be highly destructive and are like gamma rays, but with less energy.
Neutrons	Neutrons are released by nuclear weapons and in nuclear power plants. They can make nonradioactive materials radioactive.

environment. And there is no way to clean it up or dispose of it. It simply will not go away. There is only one thing that can weaken the radioactivity of nuclear wastes. Nature. Only natural decay of the wastes makes them go away, but it takes tens and hundreds of thousands of years.

Meanwhile, all those radioactive isotopes become part of the air we breathe, the food we eat, and the water we drink and use to irrigate our crops. Our bones, certain glands, and other parts of the body then store up the radioactive materials. There is no way we can get rid of them because they become part of us.

What happens when you swallow or inhale a radioactive atom? The atom may give off a burst of damaging radiation that streaks through the soft tissue of your stomach or lungs. As it does, it leaves a trail of damaged atoms behind. It can damage normal atoms by knocking off electrons. That leaves the damaged atoms with an electrical charge. The charged atoms then may violently pair up with other atoms or molecules. As they do, they may kill or injure the body cell they have entered. Weeks, months, or years later, cells injured in this way may begin to reproduce wildly. That wild behavior is what we call cancer. Radiation is a cause of cancer.

If a damaged cell happens to be a human reproductive cell, the baby born from such cells may be physically deformed. Or it may be mentally retarded. Radiation also seems to make our bodies age faster, but just how is uncertain. That is only one of the many ways we are ignorant about the dangerous effects of radiation.

Radiation and the Food Chain

Let's consider how radioactive atoms from wastes that seep into a river, for example, build up in animals and plants and eventually reach us. The radioactive isotope in our example is zinc-65, a heavy metal. For several years, a government-operated nuclear reactor near the Columbia River in Richland, Washington, has been polluting the river by leaking zinc-65 into it. This radioactive element is taken up by tiny plant organisms called plant plankton. Although the river water may contain only a small amount of the radioactive zinc, the plant plankton collect and store it. The result is that it keeps building up in their bodies.

Next, the plant plankton are gobbled up by tiny animals called animal plankton. The more plant plankton the animal plankton eat, the more zinc-65 the animal plankton store. The animal plankton may store up to ten times as much of the radioactive element in their body tissues as the plant plankton do. Small fish that eat the animal plankton may store up ten times as much of the zinc-65 as the animal plankton do. Larger fish that eat the smaller fish may have ten times as much radioactive zinc as the smaller fish do. So a large fish may end up with 10 x 10 x 10 x 10, or 10,000 times more zinc-65 than the plant plankton store. Scientists continue to study how large fish may be damaged by the radiation—and the people who eat the large fish.

The radioactive Columbia River water is used to irrigate plant crops. Plants that take up the water also collect and build up the radioactive zinc in their tissues. So do the milk cows and other grazing animals that eat the radioactive plants. People— who are the end link in the food chain—also concentrate the harmful radioactive isotope. It remains in the body, dangerously active, for fourteen years. In this way, our bodies can be exposed to high levels of radioactive isotopes. People are being harmed by such isotopes even though their concentrations when leaked into the river are so low that they can hardly be detected.

Nuclear weapons exploded in the air spray radioactive

isotopes into the environment. Scientists were puzzled several years ago when they found that Eskimos living in northern Alaska had taken in more radioactive atoms from fallout than people living near the bomb test areas. The puzzle was solved when someone thought of examining the arctic food chain. Arctic plants called lichens had been collecting and building up radioactive fallout from the air. Lichens happen to be an important food for caribou. In turn, caribou are an important food for the Eskimos. To this day the Eskimos continue to take in the radioactive atoms from nuclear weapons fallout concentrated by lichens. What long-term effects the radiation may have on the Eskimos are not known.

Our bodies take in dozens of different kinds of harmful isotopes. Among them is strontium-90, a radioactive isotope from hydrogen bomb fallout. The body mistakes the strontium for calcium and uses it to build bone tissue. The strontium has a half-life of about thirty years. Cesium-137 is another radioactive waste. The body mistakes the cesium for the beneficial element potassium and concentrates the cesium in muscle tissue. The cesium also has a half-life of about thirty years. Iodine-131 is mistaken for ordinary iodine and is taken up by the thyroid glands. Our thyroid glands help regulate body growth.

So different kinds of isotopes build up in different parts of the body. We still have much more to learn about how plants and animals store and use radioactive isotopes, all of which are harmful. Late in 1997 the National Cancer Institute released a report about radioactive fallout from nuclear weapons testing done by the U.S. in the 1950s and 1960s in a Nevada test site near Las Vegas. The report predicts that fallout from that test site may eventually cause between 7,500 and 75,000 cases of thyroid cancer.

Human activity on this planet has almost doubled the

amount of our exposure to radiation. The doubling comes mostly from medical and dental X-rays. But other sources include phosphate fertilizers, your television set, smoke detectors, and leakage from the nuclear power plants and fallout from the nuclear weapons tests mentioned above. However, there is still another source, a natural source called *background radiation*. Ever since life began on Earth, organisms have been exposed to low levels of radiation. The radiation comes from cosmic rays from space, from radioactive elements in Earth's crustal rock, such as radium and uranium, and from radioactive elements that are contained naturally in our bodies, such as potassium-40 and carbon-14. Over the many millions of years of Earth's history, plants and animals have learned to live with this low-level background radiation. In fact, the background radiation is most likely responsible for most of the gene changes (mutations) that have powered the evolution of all living things. Even so, *all* radiation is harmful.

Since long-lived radioactive wastes do not go away for tens and hundreds of thousands of years, what happens to them in the meantime? They have to be stored somewhere, and safely stored. Boiling-hot liquid radioactive nuclear wastes have to be stored in enormous holding tanks for 50 to 100 years. They cool down enough during that time so that they can be packaged and then stored deep underground. There they will remain for thousands upon thousands of years. How do we mark such deadly radioactive graveyards to warn future generations? DANGER— Keep Out? That is a problem that has not yet been solved. The U.S. Nuclear Regulatory Commission and the U.S. Environmental Protection Agency have safety regulations for waste storage for up to 10,000 years. There are two problems: What assurance do we have that earthquakes or other Earth upheavals during that time will not free and scatter the deadly wastes into the environment? And will a DANGER—Keep Out sign be readable, or even meaningful, to anyone who happens to be around 10,000 or 50,000 years from now?

The reality of such concerns reared its ugly head when the government said it wanted to store deadly radioactive wastes from nuclear power plants and wastes left over from years of making nuclear weapons. The nation's top candidate for a storage area is inside Yucca Mountain in Nevada. Some feel that the storage site to be dug deep within the mountain will be safe for the proposed 10,000 years. Others disagree.

One big safety question is about the geology of Yucca Mountain. Is it likely to be geologically stable over those many thousands of years? Or is there a good chance that earthquake and volcanic activity might pose serious threats? Since 1991 researchers have been testing how quiet or active the Yucca Mountain area is. To their surprise, a team led by geologist Brian Wernicke, of the California Institute of Technology in Pasadena, has found that ground movement is 10 to 100 times greater than

More than two billion gallons of deadly radioactive stew boils away day and night, year after year, decade after decade. The lethal wastes are stored in enormous concrete and metal holding tanks like these at government locations in Richland, Washington; Arco, Idaho; and along the Savannah River in South Carolina. There is more than enough stored radioactive stew to poison all of the water on the planet, but the exact amount is kept secret by the federal government.

expected. Wernicke suspects that the movement is being caused by the shifting of molten rock deep beneath the Yucca Mountain region. If he is right, a volcanic eruption of the mountain a few thousand years from now could blow up whatever deadly nuclear wastes might be stored there and send them flying high into the atmosphere. Once in the atmosphere, the global circulation of the air would carry thousands of tons of radioactive dust around the world.

Both the United States and Russia for decades raced to build and stockpile deadly nuclear bombs and missiles. Now they face the problem of shielding the world from those deadly weapons of mass destruction. The big question is, can they?

Tunnels cut in salt deposits a half-mile underground outside Carlsbad, New Mexico have been carved out by the Energy Department for the storage of the longest-lived, and deadliest, of radioactive wastes. The wastes are plutonium from twenty nuclear weapons plants. The plutonium will have to be kept "safe"—by someone—for 250,000 years before it becomes harmless.

Ten

Quarks, Leptons, and the God Particle

Particles of the Atomic Zoo

Matter is made of atoms. Atoms are made of electrons surrounding a nucleus, and the nucleus is made of protons and neutrons. But what are protons and neutrons made of? That question brings us to modern studies of the atom.

In their continued search for the ultimate mystery particle—the *a-tom* of Democritus—physicists shoot proton bullets into the hearts of atoms. Their "guns" are huge circular atomic raceways called *particle accelerators*. One is the Fermi National Accelerator Laboratory near Batavia, Illinois. It is a circular underground tunnel 4 miles (6.4 kilometers) around and lined with magnets. Protons are shot into the tunnel, where the magnets drive them to dizzying speeds. The protons then smash into the heart of atomic nuclei and reveal what protons and neutrons are made of.

The present picture of the atomic nucleus is a very confusing one, especially if you happen to be a nuclear physicist. It's hard to make sense of the bewildering family of particles that are the building blocks of protons and neutrons, and the forces that hold those particles together.

Our present view is that matter is made of two kinds of basic particles, called *quarks* and *leptons*. Particle physicists know of six different kinds (or "flavors") of quarks, which have been given the fanciful names up, down, strange, top, bottom, and charm. Each quark can have one of three charges, called colors, although they have nothing to do with real colors. Protons and neutrons are made up of three quarks each. A proton contains two up quarks and a down quark, each of a different "color." The quarks are held together by other particles, called gluons (as in "glue on"!). A neutron has two down quarks and one up quark. Individual quarks have never been observed, but we find them in pairs or triplets called hadrons. There are also six kinds of leptons,

A scientist at the Fermi Accelerator Laboratory in Batavia, Illinois studies tracks made by subatomic particles, in this case neutrinos. Such telltale tracks are made when particles are sped though the tunnel of a particle accelerator and made to strike a target. The resulting splash pattern of lines, curves, and spirals can then be photographed. Such photographs provide modern physicists with a means of learning more about the makeup of atoms.

QUARKS,
LEPTONS,
AND
THE
GOD
PARTICLE

including the electron and five other particles. The main difference between leptons and quarks is that leptons have no "color" charge. And there are pions, photons, positrons, muons, neutrinos, bosons, mesons, baryons, hyperons, antimatter, and more!

Antimatter is very interesting stuff. The English physicist P. A. M. Dirac first predicted its existence in 1928. Four years later the first antiparticle was discovered. It was a *positron*, or an "electron" but with a positive charge. Today we know that when two protons smash into each other forcefully, they may produce a shower of positrons, electrons, and other particles. Then, if a positron collides with an electron, or a proton collides with an antiproton, the oppositely charged particles destroy one another.

Dirac had imagined that *every* atomic particle had an antimatter twin. He further thought that antimatter particles could combine and form antiatoms. And that antiatoms could combine to form all kinds of antiobjects—antistars, antigalaxies, and even antihumans. If a human shook hands with an antihuman, both would be annihilated in a shower of gamma rays. The explosion would be powerful enough to destroy a small city.

Although we can see the work of antimatter, so far, no one has *ever* actually seen a quark or a lepton. We have looked in the sea, on the Moon, and under the bed in our search for them. But we have seen their ghosts in the giant particle accelerators when a proton is smashed and broken apart in various ways. It is like seeing the pope on television. We are seeing not the pope himself but an image of him traced out by a beam of electrons striking the inside of a glass tube painted with phosphorus. Evidence that quarks and leptons really *exist* is similarly convincing if you believe in such ghosts. And most physicists do. Add to our presently confusing picture of the atom the fact that protons and neutrons spin.

As for antimatter, scientists have observed it by the bucketful.

The end of the search for what makes an atom tick is nowhere in sight. Democritus would be amused.

Bohr and Photons

Before ending this final chapter about the ever-changing atom, we must visit one more of its architects. He is the Danish physicist Niels Bohr. In 1911 he became fascinated by Rutherford's nuclear atom. One question he asked himself was why the negatively charged electron of a hydrogen atom didn't get pulled into the nucleus of the single positively charged proton. Earlier, we promised to return to this problem. His answer was that the electron must be allowed, somehow, to move about the nucleus only in certain orbits and never fall into it or fly off. Never, that is, unless something disturbs the electron, like striking it with a cosmic hammer. He further imagined that an electron has a home orbit, called the *ground state*. Now let's substitute the words "energy level" for orbit. An atom's electron, then, has several possible energy levels that it can jump up to from its ground state, or fall down to.

So here is Bohr trying to discover the rules of nature that make invisible electrons jump farther away from or closer to an equally invisible nucleus. In the end he was able to show that if you bang an atom hard enough,

The Danish physicist Niels Bohr discovered that a hydrogen atom's electron, for example, is not anchored in an eternally fixed orbital distance from the nucleus. He said that the electron can jump to a higher orbital energy level if it absorbs energy. But it will immediately fall back to its "home" energy level and in the process give off energy. Bohr was awarded the Nobel Prize for physics in 1922. He lived from 1885 to 1962.

or heat it enough, its electrons can be made to jump up to higher energy levels. But immediately they fall back down to their "home" level, or home distance from the nucleus. When the atom is smacked, its electrons absorb the energy of the smack. But when an electron jumps back to its home level, it gives off a particle of energy called a *photon*. Never mind about all those other particles. Photons are important because when they strike our *eye we see* them as particles of light, particles of color. The size of the jump controls whether the photon will be red, blue, yellow, green, or any of the other colors we see. When you look at the red neon sign that says JOE'S DINER, all those atoms of

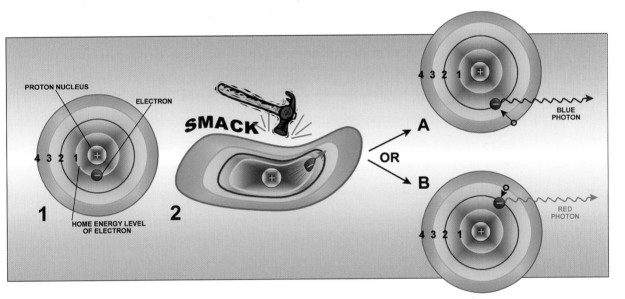

1 *A hydrogen atom is at "rest," with its single electron at its ground state energy level, or home energy level. Note there are four possible energy levels in this example.* **2** *The atom gets hit by another atom hard enough so that the electron absorbs the energy and gets bounced up to a higher energy level.* **A** *If it gets bounced up to energy level 4, it immediately jumps back down to energy level 2. As it does, it releases a tiny burst of energy, or a particle of light in the form of a blue photon.* **B** *If the electron is bounced only up to energy level 3, it immediately jumps back down to energy level 2. As it does, it emits a tiny burst of energy, a particle of light in the form of one red photon.*

neon are being energized by electricity and giving off the red photons of neon. Likewise, yellow streetlights give off the yellow photons characteristic of the sodium in a sodium vapor lamp. And the greenish streetlights give off the green photons characteristic of the mercury in a mercury vapor lamp. So each different kind of atom has photon fingerprints that identify hydrogen as hydrogen, carbon as carbon, and so on with all of the other chemical elements.

Of all the bits and pieces of atoms, it is the electron that affects our lives the most. The thousands of chemical reactions that go on inside us and around us day and night are the work of electrons joining different kinds of atoms together or driving them apart. It is the electrons of the elements sodium and chlorine that join those atoms as the compound sodium chloride ($NaCl$), common table salt. The atoms of oxygen and hydrogen are likewise bonded as that compound we cannot do without, water (H_2O).

We have learned a lot about the atom since the time of Democritus. Even so, as Mr. Romer warned earlier, "Do not think for a moment that you know the real atom." Discovery of the dozen or so nuclear particles, and the forces that control them, since the late 1800s is impressive. But we have yet to discover the basic structure of matter, the *atom* of Democritus. It still eludes our snare. Nobel Prize physicist Leon Lederman believes that there is one key mystery particle yet to be found, and when it is found we will shout Eureka! for it will make sense of the quarks, pions, and all the other particles in the atomic zoo. He believes that it will also make sense out of all matter that has ever existed in the Universe. Physicists in general call that mystery particle the Higgs boson. Lederman calls it the God Particle. "My ambition," he says, "is to live to see discovery of the God Particle reduce all of physics to a formula so elegant and simple that it will fit easily on the front of a T-shirt."

Alpha particle—a positively charged particle consisting of two protons and two neutrons; the nucleus of a helium atom.

Antimatter—a special form of matter that, if brought into contact with ordinary matter, annihilates it. Antimatter has been observed in cosmic rays and in laboratory experiments.

Atom—the smallest possible piece of an element, consisting of protons, neutrons, and electrons. Ninety-two different kinds of atoms exist in nature, ninety on Earth, two more found in stars.

Atomic number—the number of protons in the nucleus of an atom determines the atom's atomic number. For example, the element helium has two protons in its nucleus, so its atomic number is 2.

Atomic weight—the total of the number of protons and neutrons in the nucleus of an atom. For example, the element helium has two protons plus two neutrons in its nucleus, so its atomic weight is 4.

Background radiation—naturally occurring radiation from Earth materials and from space that seems to be rather constant. Such radiation comes from uranium and radon, for example.

Beta particles (rays)—streams of high-speed electrons.

Cathode—the negative electrode in a vacuum tube.

Cathode-ray tube—a common television tube, a vacuum tube in which a cathode emits electrons as a beam that strikes and illuminates a phosphorescent screen.

Cathode rays—a stream of electrons emitted by the cathode in a vacuum tube.

Chain reaction—a series of nuclear reactions in which neutrons split a uranium nucleus, which in turn releases more neutrons that split still more uranium nuclei.

Compound—substances formed by the bonding into molecules of different kinds of atoms; for example, the bonding of one atom of carbon and one atom of oxygen to form a molecule of carbon monoxide.

Cosmic rays—radiation from space consisting of protons, high-energy electrons, alpha particles, and other atomic nuclei.

Critical mass—the smallest amount of fissionable material (uranium, for example) that can be made to initiate a chain reaction and keep it going.

Electron—a negatively charged particle that is part of an atom and that is involved in the bonding of elements to form compounds.

Element—any substance made up of atoms, all of which have the same number of protons in the nucleus; for example, gold, sulfur, and oxygen are elements. An element cannot be broken down into a simpler substance by chemical means.

Fission—a nuclear reaction in which the nuclei of heavy atoms, such as uranium, are split into two more or less equal halves.

Fluorescence—light and other radiation given off by certain substances when they absorb ultraviolet radiation. Such a substance glows for as long as it is exposed to the activating radiation.

Fusion—a nuclear reaction in which nuclei join and fuse into more massive nuclei with the accompanying release of energy. For example, hydrogen nuclei in the Sun's core fuse and form nuclei of helium.

Gamma rays—very high-energy electromagnetic radiation that is highly destructive to living tissue.

Geologic time—that portion of time that took place before written history began and goes back to the formation of the planet, some 4.6 billion years ago.

Ground state—the lowest energy level of an atom.

Half-life—the length of time during which one half of the number of atoms of a radioactive element change into atoms of a different element. For example, the half-life of uranium-238 as it changes into lead-206 is 4,510 million years.

Isotope—any variety of an element whose atoms have the same atomic number but a different atomic weight; that is, the same number of protons but a different number of neutrons.

Lepton—one of the two basic kinds of particles that make up matter. There are six known kinds of leptons, one being electrons.

Molecule—the smallest piece of an element or a compound that continues to have the same chemical and physical properties.

Neutron—a nuclear particle lacking an electric charge and that has just about the same mass as a proton.

Nucleus—the central core of an atom that contains the atom's protons and neutrons.

Particle Accelerator—a circular tunnel several miles around into which subatomic particles are fired and sped to a target of atomic nuclear particles, which are shattered and so reveal their constituents, such as quarks and leptons.

Photon—a "particle" of light emitted when an electron falls from a higher energy level to a lower energy level.

Positron—the antimatter equivalent of an electron; that is, an "electron" with a positive charge.

Proton—a nuclear particle with a positive electrical charge that has just about the same mass as a neutron but is nearly two thousand times more massive than an electron.

Radioactive—An element is said to be radioactive if the atoms of the element are unstable and emit bits and pieces of themselves, called decay products

Radioactivity—the decay products of radioactive elements, such as uranium, radium, and thorium, that consist of alpha particles, electrons, and gamma rays, for example.

Quark—One of the two kinds of basic particles that make up matter. Particle physicists recognize six kinds of quarks.

Tracer isotope—a weakly radioactive isotope that can be injected into the blood stream and reveal circulation problems; they can also reveal how plants use certain nutrients and how well our bodies use what we eat.

Ultraviolet light—high-energy electromagnetic radiation, which in large doses can cause skin cancer. Ultraviolet rays are between X-rays and violet light on the electromagnetic spectrum.

X-rays—high energy radiation between gamma rays and ultraviolet rays on the electromagnetic spectrum. Since X-rays penetrate flesh but are stopped by bone tissue, they can photograph our bones.

Further Reading

Amaldi, Ginestra. *The Nature of Matter*. Chicago: University of Chicago Press, 1966.

Asimov, Isaac. *A Short History of Chemistry*. Garden City, N.Y.: Anchor Books, Doubleday, 1965.

Bernstein, Jeremy. *The Tenth Dimension*. New York: McGraw-Hill, 1989.

Gamow, George. *Thirty Years That Shook Physics*. Garden City, N.Y.: Anchor Books, Doubleday, 1966.

Greiner, Walter, and Aurel Sandulescu. "New Radioactivities." *Scientific American* (March 1990) pp: 58-67.

Lederman, Leon. *The God Particle,* with Dicjk Teresi. Boston: Houghton Mifflin, 1993.

Lenssen, Nicholas. *Nuclear Waste: The Problem That Won't Go Away*. Worldwatch Paper 106 (1991).

Monastersky, Richard. "Shifting Ground at Nuclear Waste Site." *Science News* (April 18, 1998) p. 251.

Perkins, S. "Full Report of Nuclear Test Fallout Released," *Science News* (October 11, 1997) p. 231.

Peterson, Ivars. "Proton-Go-Round." *Science News* (September 6, 1997) pp: 158-159.

Romer, Alfred. *The Restless Atom*. Garden City, N.Y.: Anchor Books, Doubleday, 1960.

Tarlé, Gregory, and Simon P. Swordy. "Cosmic Antimatter." *Scientific American* (April 1998) pp: 36-41.

Williams, Phil, and Paul N. Woessner. "The Real Threat of Nuclear Smuggling." *Scientific American* (January 1996) pp: 40-44.

Index

Page numbers for illustrations are in **boldface**.